MATH ADDITION FLASH

by Jennifer Johnson

Published by Karma Kids Learning.

0

1

2

3

4

5

$$\begin{array}{r} 1 \\ +\ 1 \\ \hline 2 \end{array}$$

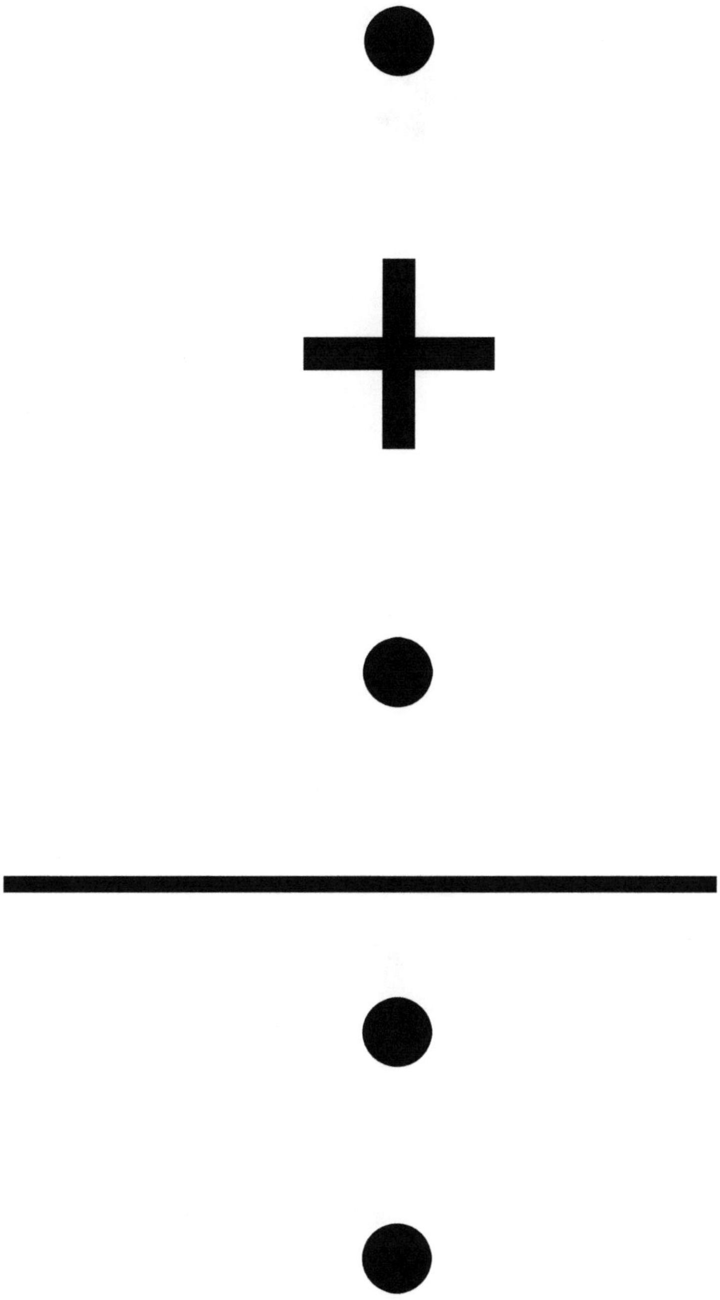

$$\begin{array}{r} 2 \\ +\ 1 \\ \hline 3 \end{array}$$

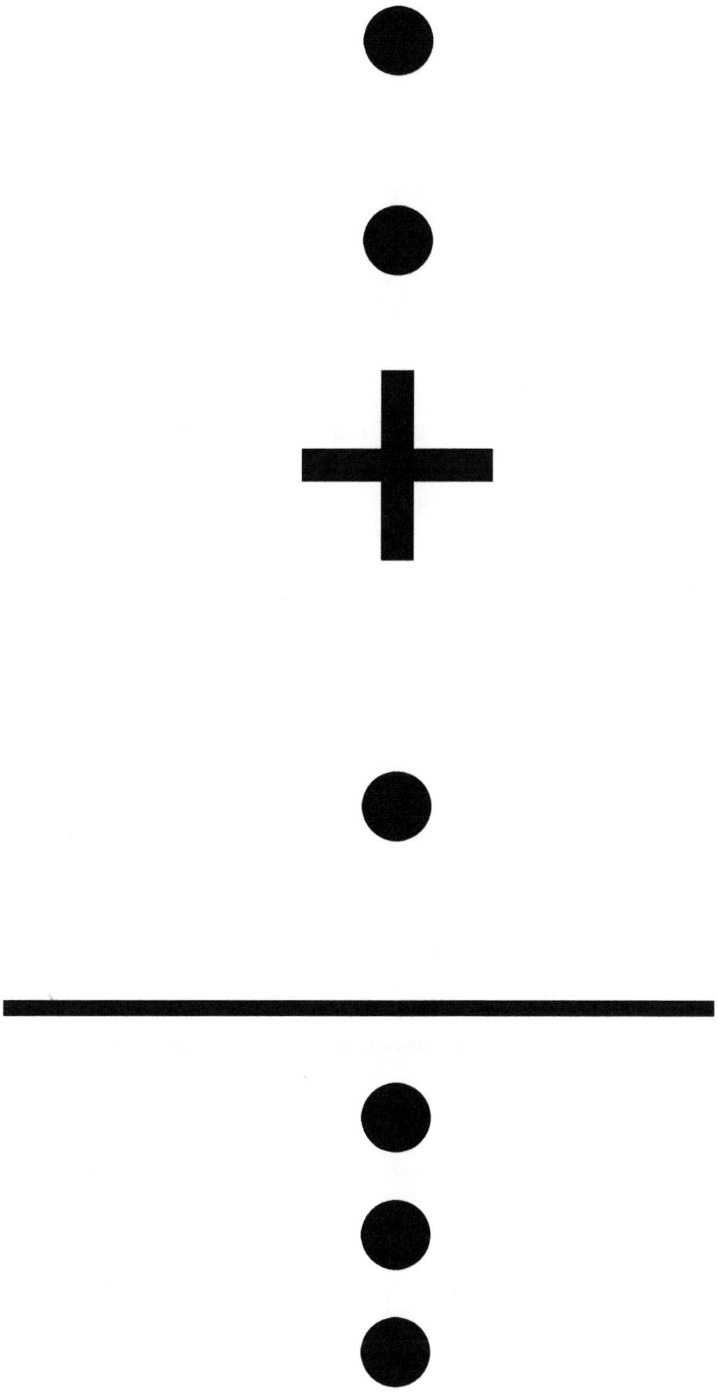

$$\begin{array}{r} 2 \\ +\ 2 \\ \hline 4 \end{array}$$

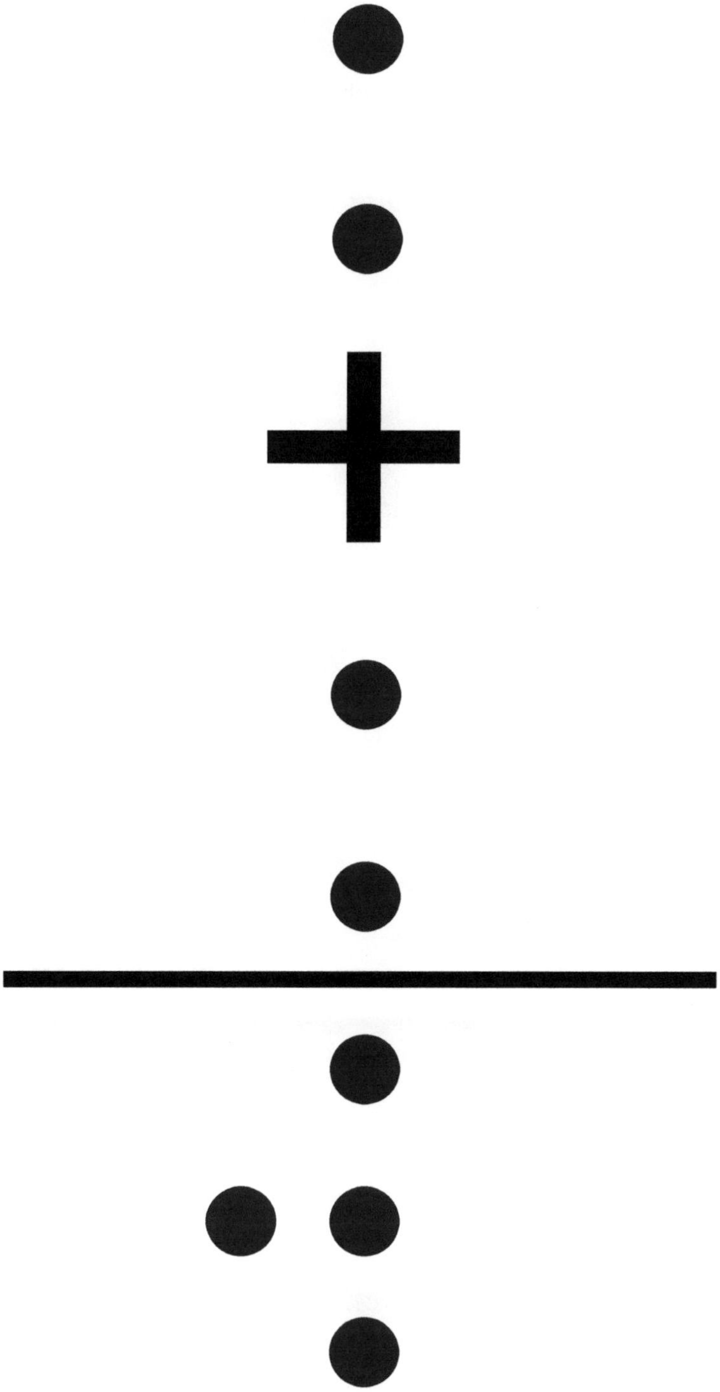

$$\begin{array}{c} 2 \\ + \\ 3 \\ \hline 5 \end{array}$$

$$\begin{array}{r} 1 \\ +\ 0 \\ \hline 1 \end{array}$$

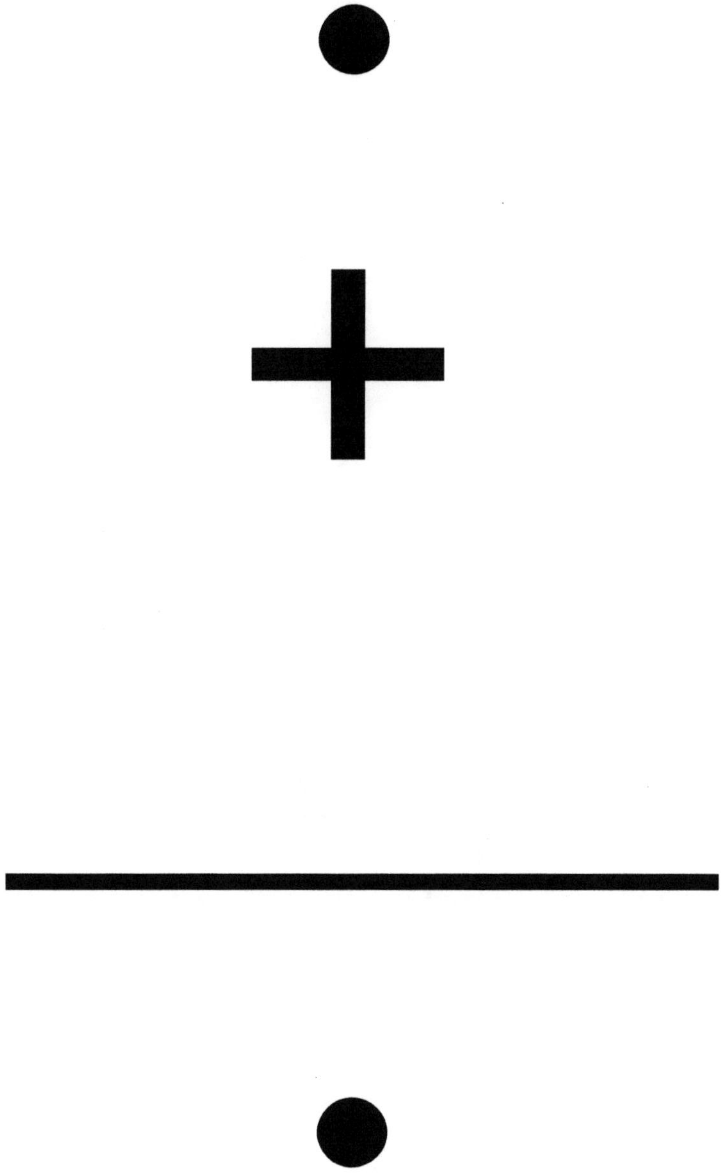

$$\begin{array}{r} 3 \\ +\ 1 \\ \hline 4 \end{array}$$

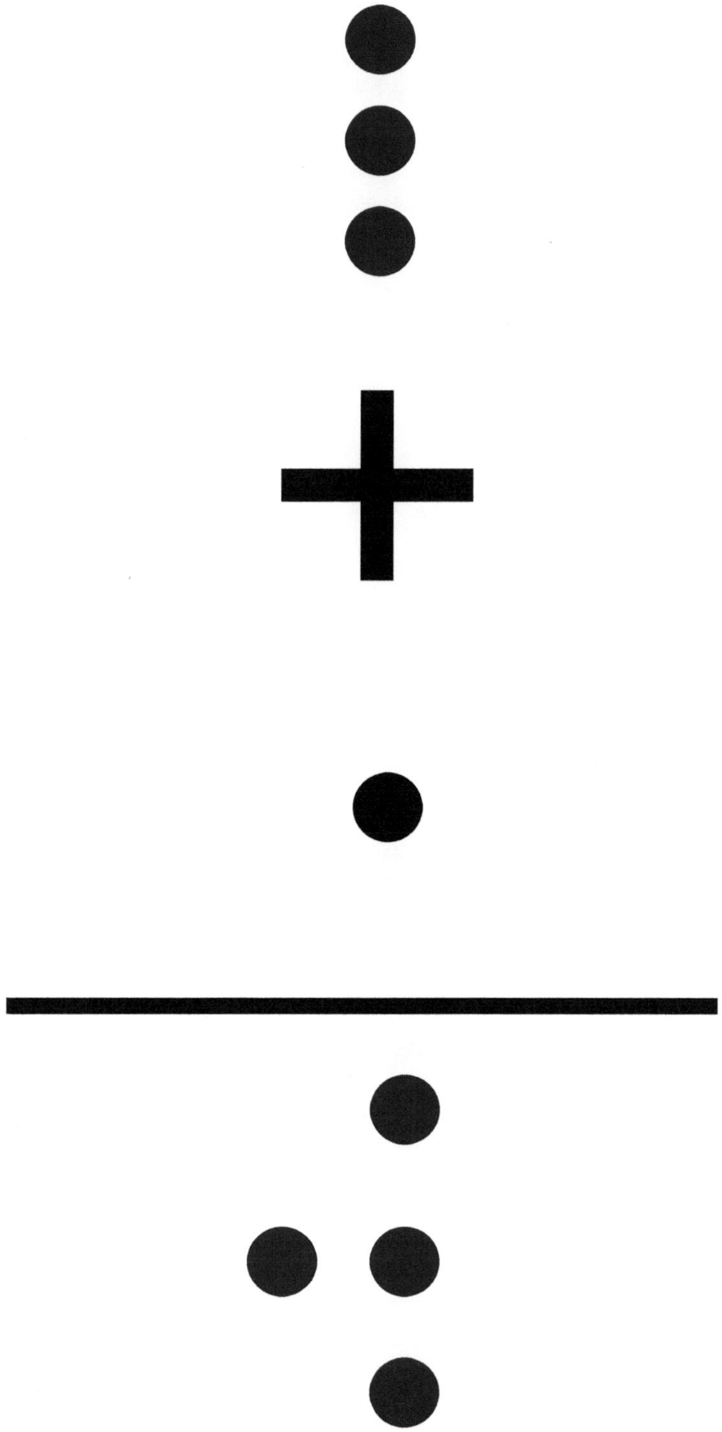

1
\+ 1

3
\+ 1

1
\+ 0

2
\+ 3

2
\+ 2

2
\+ 1

1	3	1
+	+	+
1	1	0
2	4	1

2	2	2
+	+	+
3	2	1
5	4	3

www.ingramcontent.com/pod-product-compliance
Lightning Source LLC
Chambersburg PA
CBHW050438180526
45159CB00006B/2588
9781540405241